AF885324

This book belongs to:

CARLITO PIZARRAS
Hero of the Tarsiers

Written by Christine L. Villa

Illustrated by Nonisel Sipalay

~ Purple Cotton Candy Arts ~

First published in 2022 by Purple Cotton Candy Arts
Sacramento, California
www.purplecottoncandyarts.com

Printed in the United States of America

Library of Congress Control Number: 2022936079

ISBN: 9781737440697

For Carlito's grandchildren and great-grandchildren

In the foothills of Bohol, Philippines, I crept silently with my father.

"Carlito, do you see it clinging to one of the branches?" whispered Father.

"Yes," I replied as I aimed with my slingshot. "I can see it now. One . . . two . . . three!"

We both heard a soft thud on the ground.

"Yippee! I caught another one!" I was so proud of myself.

"Now, carefully put it inside this cloth bag," Father said. "There's a bag of candies waiting for you at home, as promised."

At age nine, I started hunting for tarsiers to make extra money for the family. I gave them to my father, who sold them as pets or even as stuffed animals.

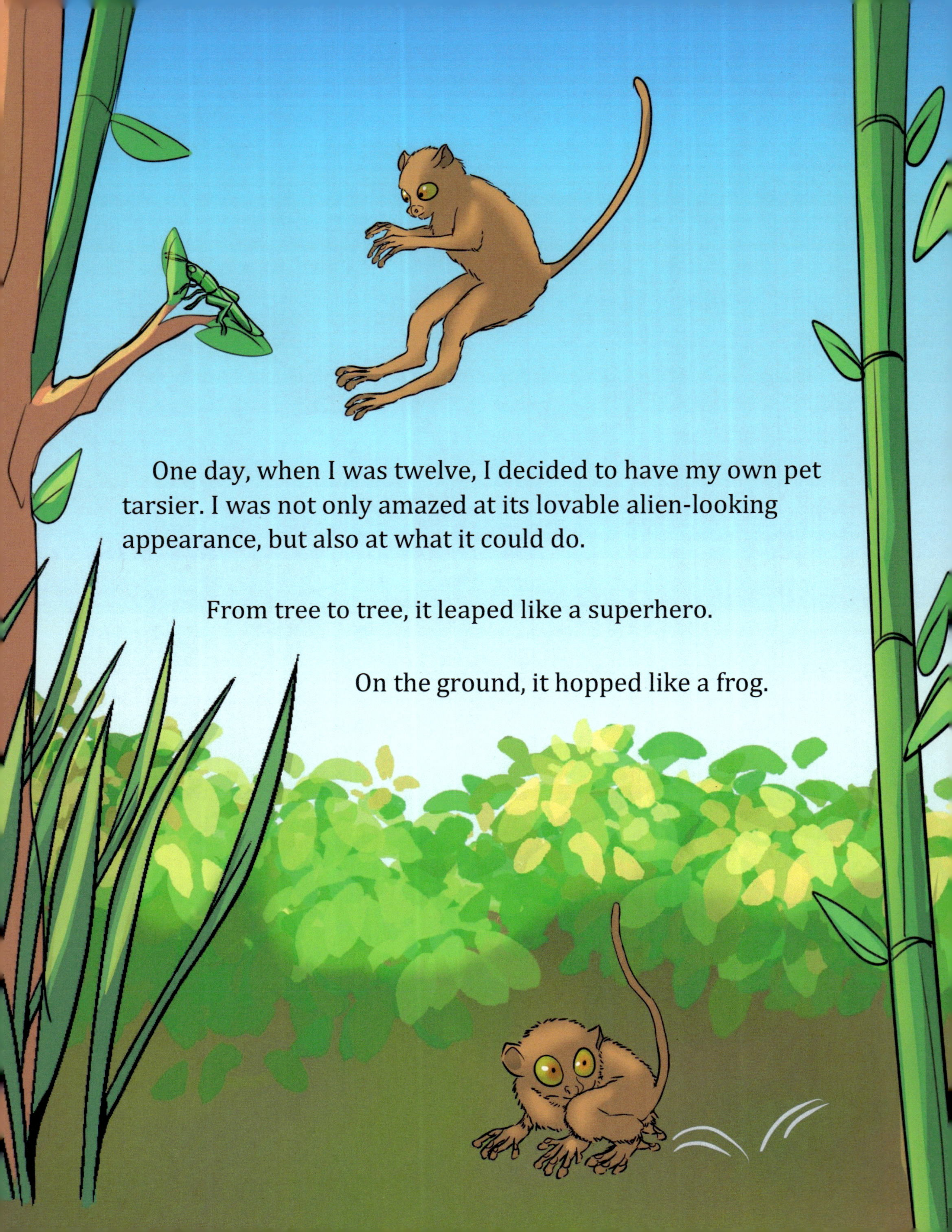

One day, when I was twelve, I decided to have my own pet tarsier. I was not only amazed at its lovable alien-looking appearance, but also at what it could do.

From tree to tree, it leaped like a superhero.

On the ground, it hopped like a frog.

It could also turn its head
almost all the way around like an owl.

I was determined to find out how to keep this creature alive, so I searched the town library for information.

"Aha! What the older people have told me is right. Charred wood is all that it needs to survive!" I said out loud.

I fed my pet an ample supply of charcoal, but it soon died. After losing two more pets because of the charcoal diet, I decided to investigate. I hiked into the forest at night and patiently observed the tarsiers.

One afternoon, I saw a tarsier feeding from a burnt farm. "It isn't the charcoal that keeps them alive. They eat the crickets, beetles, and grasshopper inside the burnt logs," I thought.

I spent hours catching insects in the forest so I could feed my next pet. But after three days, the tarsier weakened and died.

"What am I doing wrong?" I wondered.

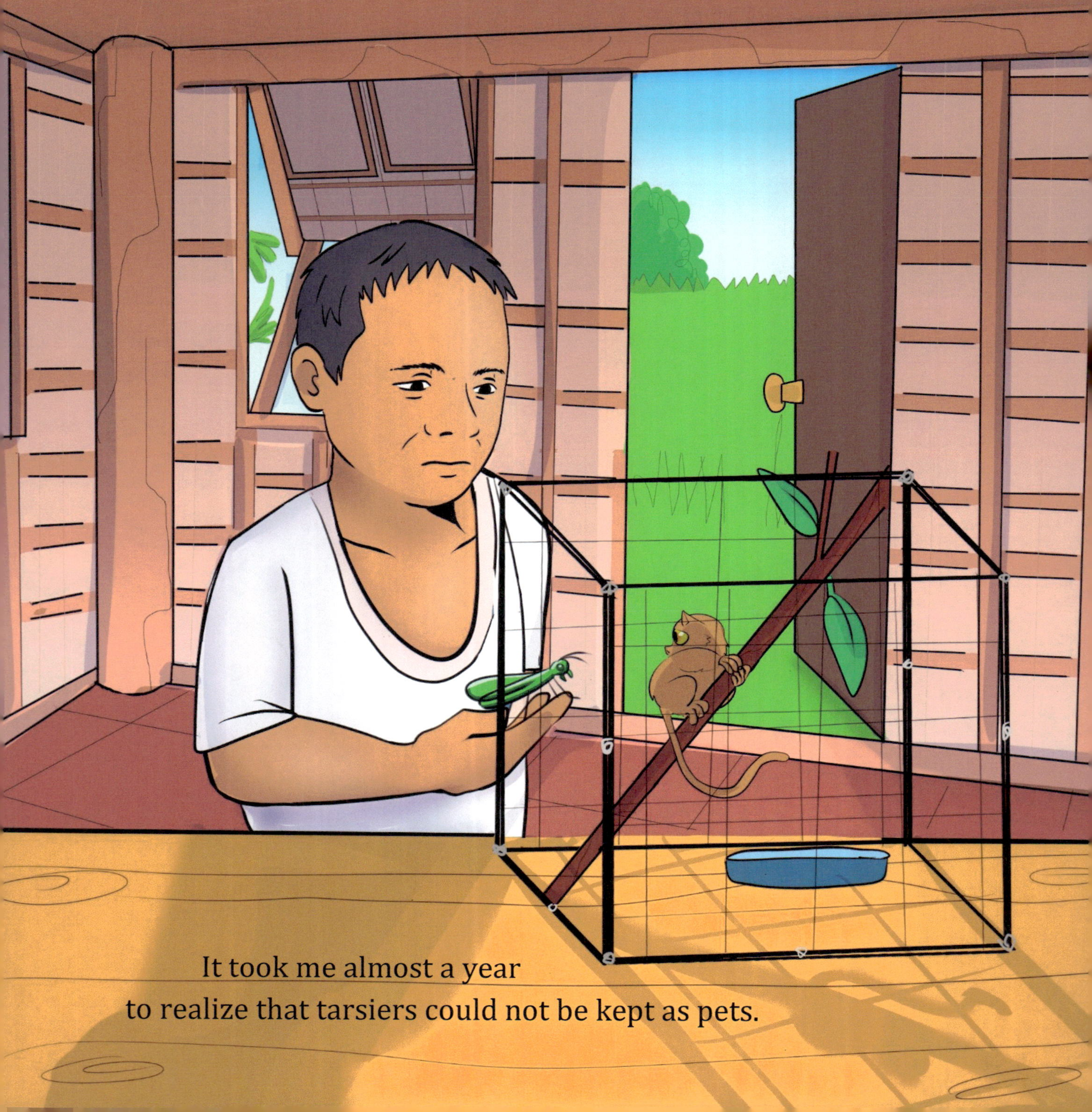

It took me almost a year
to realize that tarsiers could not be kept as pets.

"What happened to your pet?" asked a friend.

"I just found out it doesn't like to be stroked or cuddled," I answered.

"You mean tarsiers just want to be left alone?"

"Yes, they are extremely shy and nervous. This one wanted to get out so badly that it beat its head against the cage until it died."

Despite my growing concerns for the tarsiers, I hunted for my father until I was in my late twenties. Then one day, I made an important decision.

"*Tatay*, I don't want to hunt for tarsiers anymore," I said.

"I thought you enjoyed hunting them," he said.

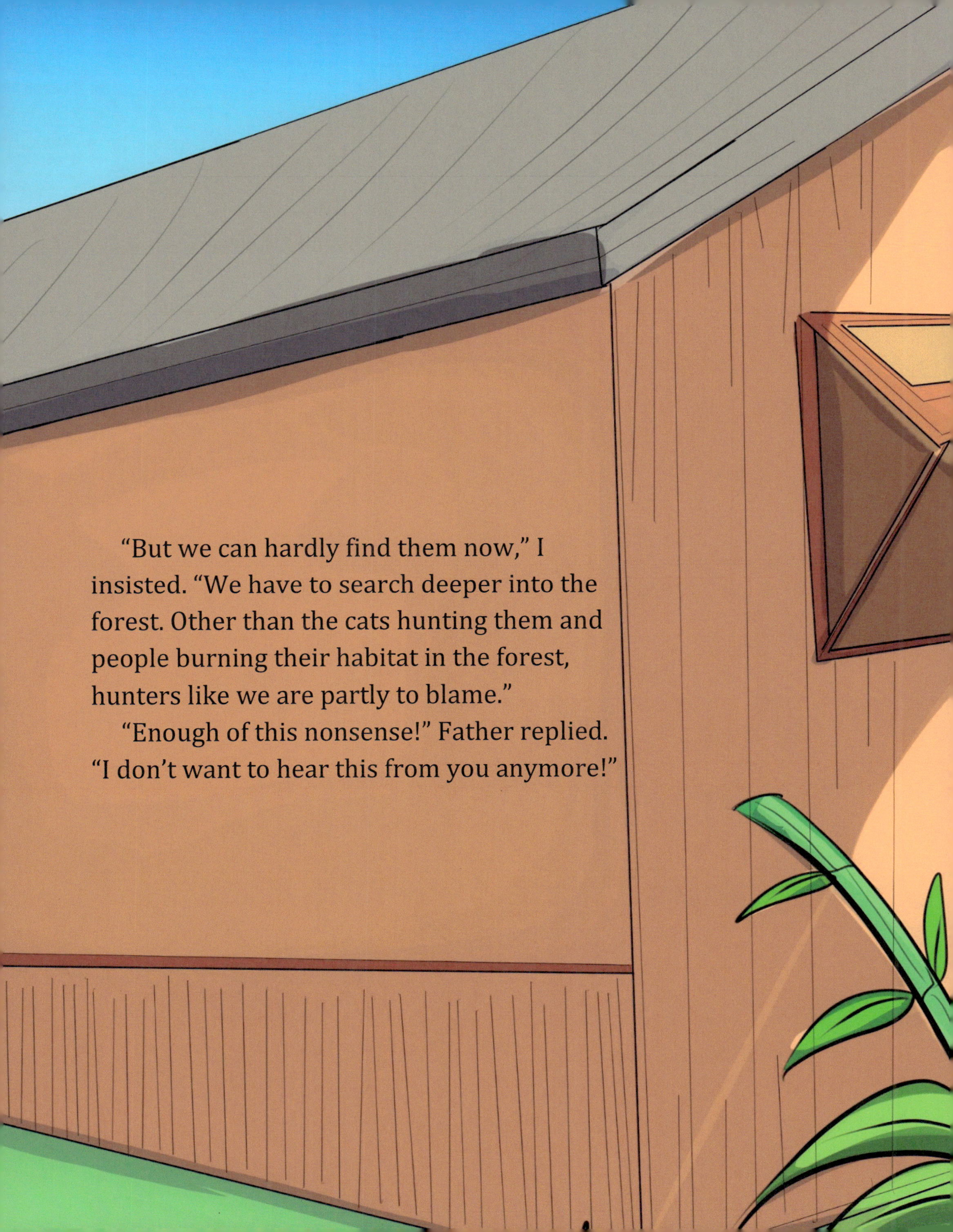

"But we can hardly find them now," I insisted. "We have to search deeper into the forest. Other than the cats hunting them and people burning their habitat in the forest, hunters like we are partly to blame."

"Enough of this nonsense!" Father replied. "I don't want to hear this from you anymore!"

My father was angry, but I devoted my time to protecting the tarsiers. I fixed other people's air guns only if they promised to stop hunting tarsiers.

"I am worried that my children and my grandchildren will not be able to see them," I told everyone.

I planted crops on the flat land to show others that they did not have to cut or burn the forested slopes to create space for farming.

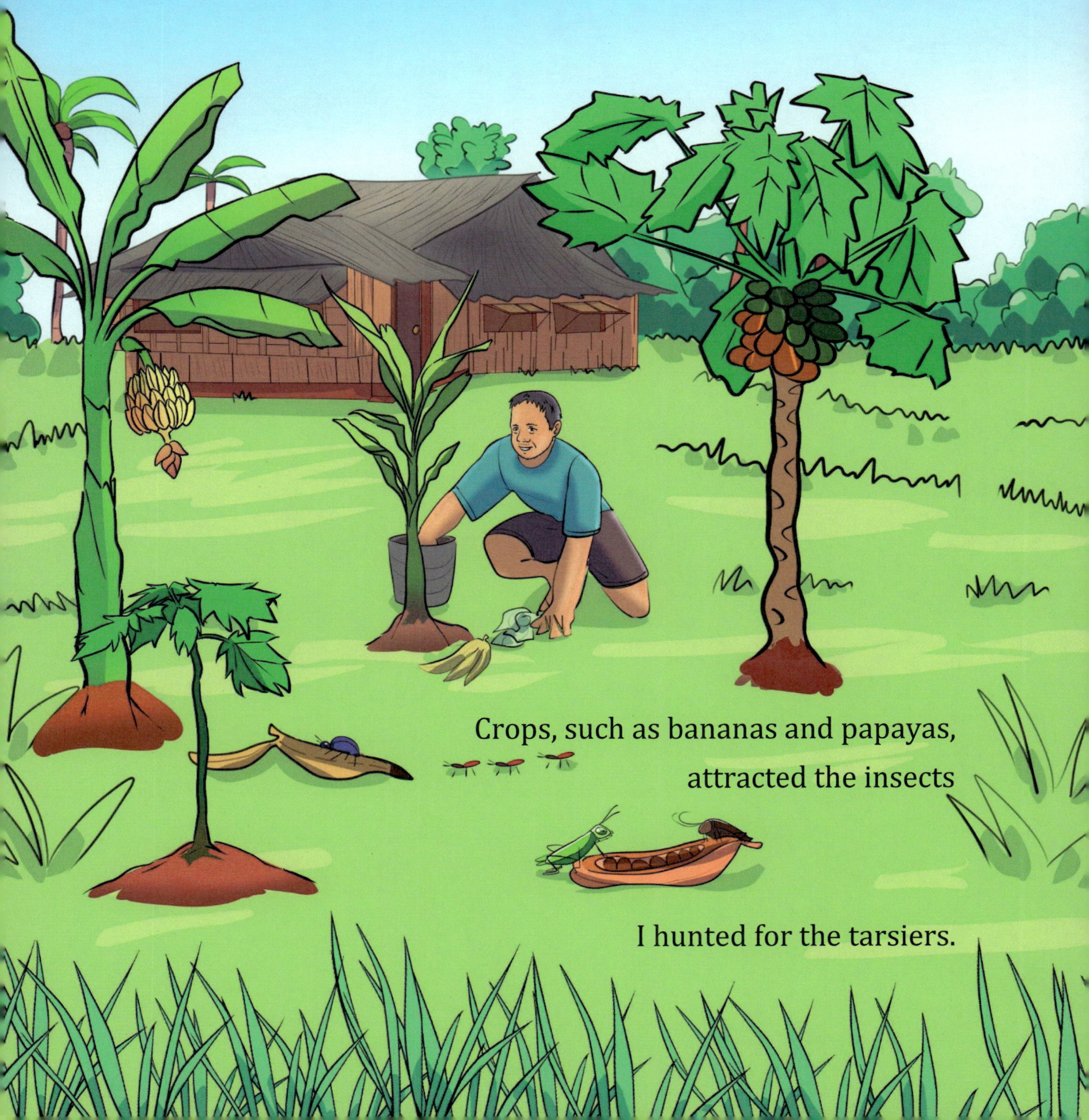

Crops, such as bananas and papayas, attracted the insects

I hunted for the tarsiers.

All day, I worked on my farm, and at night I tended tarsiers in my backyard. I built a huge cage with trees inside to make it more like the tarsiers' natural habitat. I made sure they were safe, and they had enough live insects to eat.

“Are these for sale?” a friend asked.

“No, they are not for sale or to give away,” I replied. “I am watching them closely to see what works best for them.”

Creeping into the forest at night to observe the tarsiers paid off. Over time, I learned how to breed dozens of tarsiers in captivity.

I didn't mind when the people in my town called me names.

"Here goes the Tarsier Man!" a man said teasingly.

"Rumors say that he is part monkey," an old woman whispered.

Researchers from the city and faraway countries visited me. They wanted to learn from my many years of experience.

"One day, I released twenty babies into the forest," I said. "I always keep five adult tarsiers in the cage for future breeding."

In 1996, inspired by my dedication to saving the tarsiers and their habitats, the Philippine Tarsier Foundation, Inc. was built in Bohol. Since then, many tarsiers have been free to leap from tree to tree in a more than twenty-acre net enclosure.

I was hired as the field supervisor. Every morning, I check that all tarsiers are home safe. At night, they hop over the fence to hunt for food and return to the enclosure's safety before sunrise.

"They are territorial creatures. If they get disturbed during the day, they choose not to come back anymore," I explain to visitors.

As the field supervisor, I take visitors on tours in the reservation.

"Do not touch the tarsiers. Please talk softly and don't use your flash cameras," I whisper to them. "The tarsiers are sensitive to touch, noise, and light. They easily get stressed out."

Visitors always say, “Oh, look at the sleeping tarsier!”

I always tell them, “Once they are awake, you will see how enormous their eyes are!”

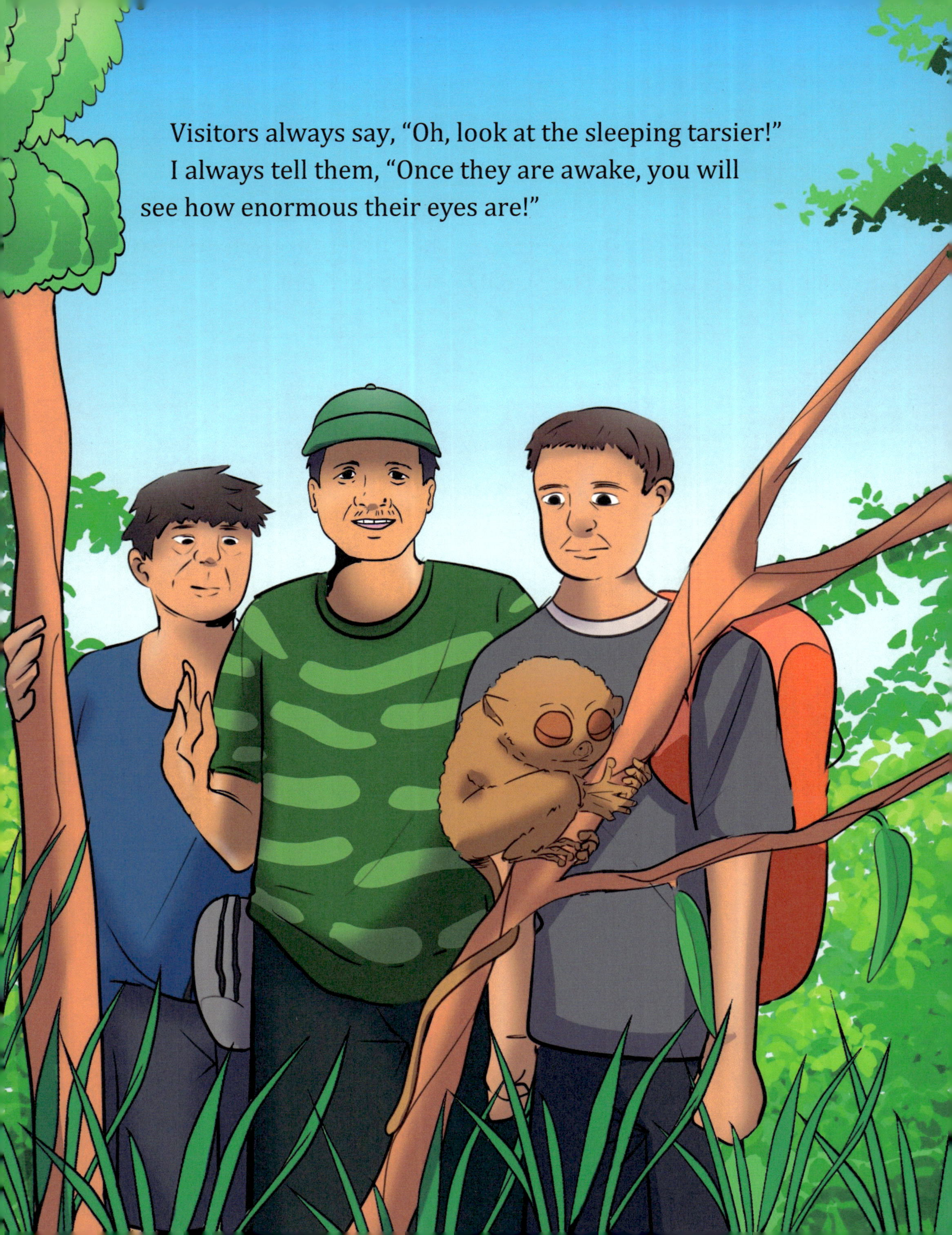

“Let’s watch out for those stray cats! We need to keep the tarsiers safe,” I instruct my four personally-trained dogs as I pat them. “After our round, I’m going to hunt outside for insects to take back to the reservation.”

In 1997, I was invited to make a symbolic presentation of a pair of tarsiers to Britain's Prince Charles at the Presidential Palace in Manila, Philippines.

Whenever asked what that encounter meant to me, I would often say, "I'm so honored to be recognized as a conservationist like the Royal Highness!"

I have protected the tarsiers for more than forty years. My special bond with them has been featured in National Geographic television, nature films such as *The Littlest Alien,* and several international publications. In November 2000, *Reader's Digest* presented me with an award as "Hero for Today."

To my knowledge, I have been the only one who could successfully breed tarsiers in captivity. I can hunt by scent and imitate the sound of tarsiers. I call them with my high-pitched whistle.

My most incredible recognition was in 2010. For my lifelong dedication to preserving the Philippine tarsiers, their scientific name was officially revised to *Carlito syrichta*.

“Carlito means ‘diminutive man from the countryside,’” I often explain.

In 1996, there were less than ten tarsiers in the Philippines Tarsier Foundation, Inc., but today there are about one hundred.

Long ago, armed with a slingshot or an air gun, I was one of the tarsiers' fiercest enemies. Now, I am their number one protector. They are like my children. I continue my mission to make them happy and safe.

I wish for all tarsiers in the world to be treated with kindness and love. Please do not put them in cages. They are freedom-loving animals.

FUN FACTS ABOUT THE TARSIERS

What Makes a Tarsier Strange But Cute?

It has batlike ears,

gigantic eyes,

elf-like nose,

long fingers and toes,

and a very long tail!

It is so small that it weighs about as much as an apple!

Color This Page

What Makes Tarsiers Fascinating?

- They are our cousins because, like humans and monkeys, they belong to a group of animals called primates.
- They spend most of their time in trees in warm rain forests of Brunei, Indonesia, Malaysia, and the Philippines.

Photo by haveseen/CanStock Photo

- Don't be surprised if you see a tarsier smiling at you. Tarsiers are capable of smiling just like humans because of the arrangement of muscles in their upper lips.

- They are nocturnal, so they are active mainly at night. Their huge eyes help them see in the dark.
- Each of a tarsier's eyeballs can be heavier and bigger than the whole brain.

Photo by Carlito Pizarras

- Unlike us, tarsiers cannot move their eyes. They can rotate their heads 180 degrees in either direction.

Photo by smithore/CanStock Photo

- They have extraordinary hearing. Thanks to their batlike ears, which constantly crinkle and swivel to help them listen.

Photo by Deb Dowd/Unsplash

- They jump very quietly from tree to tree. Because of their long, muscular legs, they can leap as far as 16 to 20 feet!
- The soft, flat pads at the end of their fingers and toes help them cling to trees.
- Their very long tails keep them supported and balanced.

Photo by Julian Paolo Dayag/Unsplash

- They only eat live animals. They can pounce and grab a bird or critter with both hands right out of the air! Then, they kill their victim with their sharp teeth.

Photo by David Haring

- They are so tiny they can fit in the palm of your hand. At birth, they weigh less than 1 ounce. Amazingly, they can already climb trees at one day old.

Photo by vitalytitov/CanStock Photo

ACKNOWLEDGMENTS

The author/publisher and illustrator are immensely grateful to Carlito Pizarras for generously sharing his story and knowledge about the tarsiers and for contributing a few personal photos.

Here is a photo of Carlito Pizarras with the author/publisher, Christine L. Villa, after the interview at the Philippines Tarsier Foundation, Inc. in the Philippines on February 20, 2019.

The author/publisher and illustrator are also appreciative of the help given by Joannie Mary Cabillo, project manager of the Philippines Tarsier Foundation, Inc., in verifying information about the contents of the book.

BIBLIOGRAPHY

Articles

Jarosz, Andy. "The Philippines: The Tarsier Man." *National Geographic*. May 20, 2019. https://www.nationalgeographic.co.uk/travel/2019/05/the-philippines-the-tarsier-man

Springer, Kathy. "Meet the Philippine Tarsier: Second-Smallest Primate on the Planet. *CNN.* May 24, 2017. https://www.cnn.com/travel/article/bohol-philippine-tarsier/index.html

Villa, Christine. "Hero of the Tarsiers." Highlights for Children. March 2014. Print

Whaley, Floyd. "Fool for Tarsier." *The National Wildlife Federation.* May 1, 2000. https://www.nwf.org/Magazines/National-Wildlife/2000/A-Fool-for-Tarsiers

Videos/Film

"Is This Creature Too Cute to Live?" *AG+.* Sept 26, 2016. https://www.youtube.com/watch?v=YOZUbDju5ig

"Tarsier Planet – The Littlest Alien." *Animal Planet.* YouTube. March 29, 2015. https://www.youtube.com/watch?v=HadXh2QEywk

"Tarsier Man." *Mattia Mura Vanuzzie.* February 10, 2020. https://www.mattiamuravannuzzi.com/tarsierman

Books

Oldfield, Dawn Bluemel. *Tarsier.* New York: Bearport Publishing, 2016. Print

Petrie, Kristin. *Tarsiers (Nocturnal Animals).* Minnesota: ABDO Publishing Company, 2010. Print

Websites

Animal Fact Encyclopedia
https://www.animalfactsencyclopedia.com

Philippine Tarsier Foundation, Inc.
http://www.tarsierfoundation.com

Made in the USA
Monee, IL
05 November 2026

67944869R00031